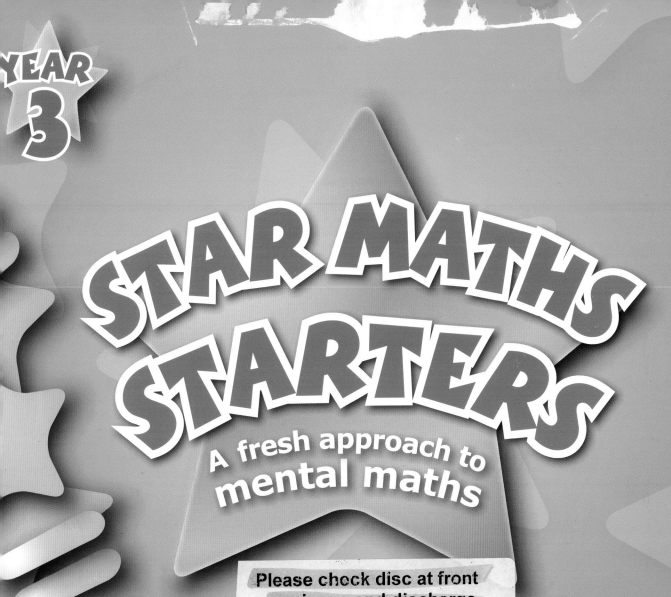

YEAR 3

STAR MATHS STARTERS

A fresh approach to mental maths

Please check disc at front on issue and discharge

TERMS AND CONDITIONS

IMPORTANT - PERMITTED USE AND WARNINGS - READ CAREFULLY BEFORE USING

Minimum specification:
- PC with a CD-ROM drive and 512 Mb RAM (recommended)
- Windows 98SE or above/Mac OSX.1 or above
- Recommended minimum processor speed: 1 GHz
- Facilities for printing

ogill and Anthony David

D1424882

3/2011

Authors
Julie Cogill and Anthony David

Anthony David dedicates this book to his wife Peachey, and sons Oliver and Samuel.

Editors
Niamh O'Carroll and Kate Pedlar

Assistant Editor
Margaret Eaton

Illustrator
Theresa Tibbetts (Beehive Illustration)

Series Designer
Joy Monkhouse

Designers
Micky Pledge and Melissa Leeke

Text © 2008 Julie Cogill and Anthony David
© 2008 Scholastic Ltd

CD-ROM development in association with Vivid Interactive

Designed using Adobe InDesign and Adobe Illustrator

Published by Scholastic Ltd
Book End, Range Road, Witney,
Oxfordshire OX29 0YD
www.scholastic.co.uk

Printed by Bell & Bain Ltd, Glasgow
2 3 4 5 6 7 8 9   0 1 2 3 4 5 6 7

ISBN 978-1407-10009-8

Mixed Sources
Product group from well-managed forests and other controlled sources
www.fsc.org Cert no. TT-COC-002769
© 1996 Forest Stewardship Council
FSC

ACKNOWLEDGEMENTS
Extracts from the Primary National Strategy's *Primary Framework for Mathematics* (2006) www.standards.dfes.gov.uk/primaryframework, *Renewing the Primary Framework* (2006) and the Interactive Teaching Programs originally developed for the National Numeracy Strategy © Crown copyright. Reproduced under the terms of the Click Use Licence.

Every effort has been made to trace copyright holders for the works reproduced in this book, and the publishers apologise for any inadvertent omissions.

British Library Cataloguing-in-Publication Data
A catalogue record for this book is available from the British Library.

Introduction

In the 1999 *Framework for Teaching Mathematics* the first part of the daily mathematics lesson is described as 'whole-class work to rehearse, sharpen and develop mental and oral skills'. The Framework identified a number of short, focused activities that might form part of this oral and mental work. Teachers responded very positively to these 'starters' and they were often judged by Ofsted to be the strongest part of mathematics lessons.

However, the renewed *Primary Framework for Mathematics* (2006) highlights that the initial focus of 'starters', as rehearsing mental and oral skills, has expanded to become a vehicle for teaching a range of mathematics. 'Too often the "starter" has become an activity extended beyond the recommended five to ten minutes' (*Renewing the Primary Framework for mathematics: Guidance paper,* 2006). The renewed Framework also suggests that 'the focus on oral and mental calculation has been lost and needs to be reinvigorated'.

Star Maths Starters aims to 'freshen up' the oral and mental starter by providing focused activities that help to secure children's knowledge and sharpen their oral and mental skills. It is a new series, designed to provide classes and teachers with a bank of stimulating interactive whiteboard resources for use as starter activities. Each of the 30 starters offers a short, focused activity designed for the first five to ten minutes of the daily mathematics lesson. Equally, the starters can be used as stand-alone oral and mental maths 'games' to get the most from a spare ten minutes in the day.

About the book

Each book includes a bank of teachers' notes linked to the interactive whole-class activities on the CD-ROM. A range of additional support is also provided, including planning grids, classroom resources, generic support for using the interactive whiteboard in mathematics lessons, and an objectives grid.

Objectives grid

A comprehensive two-page planning grid identifies links to the *Primary Framework for Mathematics* strands and objectives. The grid also identifies one of six starter types, appropriate to each interactive activity (see page 7 for further information).

Starter Number	Star Starter Title	Page No.	Strand	Learning objective as taken from the Primary Framework for Mathematics	Type of Starter
16	Maths Boggle: money problems	28	Knowing and using number facts	Use knowledge of number operations and corresponding inverses, including doubling and halving, to estimate and check calculations	Refine
17	Function machine: addition and subtraction	29	Calculating	Add or subtract mentally combinations of one-digit and two-digit numbers	Reason
18	Targets: addition and subtraction	30	Calculating	Add or subtract mentally combinations of one-digit and two-digit numbers	Rehearse
19	Maths Boggle: addition and subtraction	31	Calculating	Add or subtract mentally combinations of one-digit and two-digit numbers	Refine
20	Number line (ITP)	32	Calculating	Add or subtract mentally combinations of one-digit and two-digit numbers	Refine
21	Function machine: multiplication and division	33	Calculating	Use practical and informal written methods to multiply and divide two-digit numbers	Reason
22	Symmetry (ITP)	34	Understanding shape	Draw the reflection of a shape in a mirror line along one side	Refine
23	Maps and directions	35	Understanding shape	Read and record the vocabulary of position, direction and movement, using the four compass directions to describe movement about a grid	Rehearse

Highlighted text indicates the end-of-year objectives

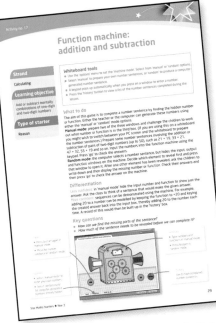

Activity pages

Each page of teachers' notes includes:

Learning objectives
Covering the strands and objectives of the renewed *Primary Framework for Mathematics*

Type of starter
Identifying one or more of the 'six Rs' of oral and mental work (see page 7)

Whiteboard tools
Identifying the key functions of the accompanying CD-ROM activity

What to do
Outline notes on how to administer the activity with the whole class

Differentiation
Adapting the activity for more or less confident learners

Key questions
Probing questions to stimulate and sustain the oral and mental work

Annotations
At-a-glance instructions for using the CD-ROM activity.

Whiteboard hints and tips

Each title offers some general support identifying practical mathematical activities that can be performed on any interactive whiteboard (see pages 8–9).

Recording sheets

Two recording sheets have been included to support your planning:
- Planning for the six Rs: plan a balance of activities across the six Rs of mental and oral maths (see page 7).
- Star Maths Starters diary: build a record of the starters used (titles, objectives covered, how they were used and dates they were used).

About the CD-ROM

Types of activity

Each CD-ROM contains 30 interactive starter activities for use on any interactive whiteboard. These include:

Interactive whiteboard resources
A set of engaging interactive activities specifically designed for *Star Maths Starters*. The teachers' notes on pages 13–42 of this book explain how each activity can be used for a ten-minute mental maths starter, with annotated screen shots giving you at-a-glance support. Similarly, a 'what to do' function within each activity provides at-the-board support.

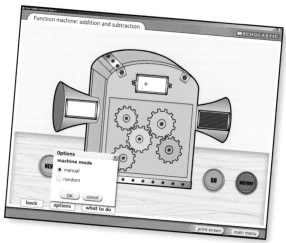

Interactive teaching programs (ITPs)

A small number of ITPs, originally developed by the National Numeracy Strategy, has been included on each CD-ROM. They are simple programs that model a range of objectives, such as data presentation or fraction bars. Their strength is that they are easy to read and use. If you press the Esc button the ITP will reduce to a window on the computer screen. It can then be enlarged or more ITPs can be launched and set up to model further objectives, or simply to extend the objective from that starter. To view the relevant 'what to do' notes once an ITP is open, press the Esc button to gain access to the function on the opening screen of the activity.

Interactive 'notepad'

A pop-up 'notepad' is built into a variety of activities. This allows the user to write answers or keep a record of workings out and includes 'pen', 'eraser' and 'clear' tools.

Teacher zone

This teachers' section includes links from the interactive activities to the *Primary Framework for Mathematics* strands, together with editable objectives grids, planning grids and printable versions of the activity sheets on pages 43–46.

How to use the CD-ROM

System requirements

Minimum specification
- PC with a CD-ROM drive and 512 Mb RAM (recommended)
- Windows 98SE or above/Mac OSX.1 or above
- Recommended minimum processor speed: 1 GHz

Getting started

The *Star Maths Starters* CD-ROM should auto run when inserted into your CD drive. If it does not, use **My Computer** to browse the contents of the CD-ROM and click on the 'Star Maths Starters' icon.

From the start-up screen you will find four options: select **Credits** to view a list of credits. Click on **Register** to register the product to receive product updates and special offers. Click on **How to use** to access support notes for using the CD-ROM. Finally, if you agree to the terms and conditions, select **Start** to move to the main menu.

For all technical support queries, please phone Scholastic Customer Services on 0845 6039091.

The six Rs of oral and mental work

In the guidance paper *Renewing the Primary Framework for mathematics* (2006), the Primary National Strategy identified six features of children's mathematical learning that oral and mental work can support. The description of the learning and an outline of possible activities are given below:

Six Rs	Learning focus	Possible activities
Rehearse	To practise and consolidate existing skills, usually mental calculation skills, set in a context to involve children in problem solving through the use and application of these skills; use of vocabulary and language of number, properties of shapes or describing and reasoning.	Interpret words such as more, less, sum, altogether, difference, subtract; find missing numbers or missing angles on a straight line; say the number of days in four weeks or the number of 5p coins that make up 35p; describe part-revealed shapes, hidden solids; describe patterns or relationships; explain decisions or why something meets criteria.
Recall	To secure knowledge of facts, usually number facts; build up speed and accuracy; recall quickly names and properties of shapes, units of measure or types of charts, graphs to represent data.	Count on and back in steps of constant size; recite the 6-times table and derive associated division facts; name a shape with five sides or a solid with five flat faces; list properties of cuboids; state units of time and their relationships.
Refresh	To draw on and revisit previous learning; to assess, review and strengthen children's previously acquired knowledge and skills relevant to later learning; return to aspects of mathematics with which the children have had difficulty; draw out key points from learning.	Refresh multiplication facts or properties of shapes and associated vocabulary; find factor pairs for given multiples; return to earlier work on identifying fractional parts of given shapes; locate shapes in a grid as preparation for lesson on coordinates; refer to general cases and identify new cases.
Refine	To sharpen methods and procedures; explain strategies and solutions; extend ideas and develop and deepen the children's knowledge; reinforce their understanding of key concepts; build on earlier learning so that strategies and techniques become more efficient and precise.	Find differences between two two-digit numbers, extend to three-digit numbers to develop skill; find 10% of quantities, then 5% and 20% by halving and doubling; use audible and quiet counting techniques to extend skills; give coordinates of shapes in different orientations to hone concept; review informal calculation strategies.
Read	To use mathematical vocabulary and interpret images, diagrams and symbols correctly; read number sentences and provide equivalents; describe and explain diagrams and features involving scales, tables or graphs; identify shapes from a list of their properties; read and interpret word problems and puzzles; create their own problems and lines of enquiry.	Tell a story using an interactive bar chart, altering the chart for children to retell the story; starting with a number sentence (eg 2 + 11 = 13), children generate and read equivalent statements for 13; read values on scales with different intervals; read information about a shape and eliminate possible shapes; set number sentences in given contexts; read others' results and offer new questions and ideas for enquiry.
Reason	To use and apply acquired knowledge, skills and understanding; make informed choices and decisions, predict and hypothesise; use deductive reasoning to eliminate or conclude; provide examples that satisfy a condition always, sometimes or never and say why.	Sort shapes into groups and give reasons for selection; discuss why alternative methods of calculation work and when to use them; decide what calculation to do in a problem and explain the choice; deduce a solid from a 2D picture; use fractions to express proportions; draw conclusions from given statements to solve puzzles.

Each one of the styles of starter enables children to access different mathematical skills and each has a different outcome, as identified above. A bingo game, for example, provides a good way of rehearsing number facts, whereas a 'scales' activity supports reading skills. In the objectives grid on pages 10-11, the type of each Star Maths Starters activity is identified to make it easier to choose appropriate styles of starter matched to a particular objective. A 'Six Rs' recording sheet has also been provided on page 12 (with an editable version on the CD-ROM) to track the types of starter you will be using against the strands of the renewed Framework.

Using the interactive whiteboard in primary mathematics

The interactive whiteboard is an invaluable tool for teaching and learning mathematics. It can be used to demonstrate and model mathematical concepts to the whole class, offering the potential to share children's learning experiences. It gives access to powerful resources – audio, video, images, websites and interactive activities – to discuss, interact with and learn from. *Star Maths Starters* provides 30 quality interactive resources that are easy to set up and use and which help children to improve their mathematical development and thinking skills through their use as short, focused oral and mental starters.

Whiteboard resources and children's learning

There are many reasons why the whiteboard, especially in mathematics, enhances children's learning:

- Using high-quality interactive maths resources will engage children in the process of learning and developing their mathematical thinking skills. Resources such as maths games can create a real sense of theatre in the whole class and promote a real desire to achieve and succeed in a task.
- As mentioned above, the whiteboard can be used to demonstrate some very important mathematical concepts. For example, many teachers find that children understand place value much faster and more thoroughly through using interactive resources on a whiteboard. Similarly, the whiteboard can support children's visualisation of mathematics, especially for 'Shape and Space' activities.
- Although mathematics usually has a correct or incorrect answer, there are often several ways of reaching the same result. The whiteboard allows the teacher to demonstrate methods and encourages children to present and compare their own mental or written methods of calculation.

Using a whiteboard in Year 3

An interactive whiteboard can be used for a variety of purposes in Year 3 mathematics lessons. These include:

- recording estimates and actual positions of numbers on a number line – then using whiteboard software to compare differences between the estimates and the actual numbers;
- using interactive multiplication squares to practise and check children's knowledge of times tables and related division facts;
- identifying and estimating fractions of shapes and using diagrams to compare fractions and establish equivalents;
- using squared paper on the whiteboard to prepare a range of 2D shapes, pictures and patterns for the children to describe;
- using software programs to collect and organise data into graphs and charts, then using the whiteboard to display and interpret the data.

Practical considerations

For the teacher, the whiteboard has the potential to save preparation and classroom time, as well as providing more flexible teaching.

ICT resources for the interactive whiteboard often involve numbers that are randomly generated, so that possible questions or calculations stemming from a single resource may be many and varied. This enables resources to be used for a longer or shorter time period depending on the purpose of the activity and how children's learning is progressing. *Star Maths Starters* includes many activities of this type.

From the very practical point of view of saving teachers' time, particularly in the starter activity, it is often easier to set up mathematics resources more quickly than those for other subjects. Once the software is familiar, preparation time is saved especially when there is need for clear presentation, as in drawing shapes accurately or creating charts and diagrams for 'Handling data' activities.

Maths resources on the interactive whiteboard are often flexible and enable differentiation so that a teacher can access different degrees of difficulty using the same software. Last but not least, whiteboard resources save time writing on the board and software often checks calculations, if required, which enables more time both for teaching and assessing children's understanding.

Using *Star Maths Starters* interactively

Much has been said and written about interactivity in the classroom but it is not always clear what this means. For example, children coming out to the board and ticking a box is not what is meant by 'whole-class interactive teaching and learning'. In mathematics it is about challenging children's ideas so that they develop their own thinking skills and, when appropriate, encouraging them to make connections across different mathematical topics. As a teacher, this means asking suitable questions and encouraging children to explore and discuss their methods of calculation and whether there are alternative ways of achieving the same result. *Star Maths Starters* provides some examples of key questions that could be asked while the activities are being undertaken, together with suggestions for how to engage less confident learners and stretch the more confident.

If you already have some experience in using the whiteboard interactively then we hope the teaching suggestions set out in this book will take you further. What is especially important is the facility the whiteboard provides to share pupils' mathematical learning experiences. This does not mean just asking children to suggest answers, but using the facility of the board to display and discuss ideas so that everyone can share in the learning experience. Obviously, this needs to be in a way that explores and relates the thinking of individuals to the context of the learning that is happening.

In the best whiteboard classrooms, teachers comment that the board provides a shared learning experience between the teacher and the class, in so far as the teacher may sometimes stand aside while children themselves are discussing their own mathematical methods and ideas.

Starter Number	Star Starter Title	Page No.	Strand	Learning objective as taken from the Primary Framework for Mathematics	Type of Starter
1	Grouping (ITP)	13	Using and applying mathematics	Solve one-step and two-step problems involving numbers	Refine
2	Shopping: using money	14	Using and applying mathematics	Solve one-step problems involving money	Refine
3	Twenty cards (ITP): largest and smallest numbers	15	Counting and understanding number	Read, write and order whole numbers to at least 1000	Reason
4	Bricks: ordering whole numbers	16	Counting and understanding number	Read, write and order whole numbers to at least 1000	Reason
5	Pirates: adding multiples of 10	17	Counting and understanding number	Count on from and back to zero in multiples of 10	Recall
6	Number square: counting forwards and backwards	18	Counting and understanding number	Count on, from and back to zero	Refresh
7	Sorting machine: numbers	19	Counting and understanding number	Round two-digit or three-digit numbers to the nearest 10 or 100	Refresh
8	Dominoes: fractions of shapes	20	Counting and understanding number	Identify and estimate fractions of shapes	Reason
9	Fractions (ITP)	21	Counting and understanding number	Use diagrams to compare fractions and establish equivalents	Refine
10	Number sentence builder: find the missing number	22	Knowing and using number facts	Derive and recall all addition and subtraction facts for each number to 20	Reason
11	Number spinners (ITP)	23	Knowing and using number facts	Derive and recall all addition and subtraction facts for each number to 20	Refine
12	Bingo: multiples of 2, 5 and 10	24	Knowing and using number facts	Recognise multiples of 2, 5 or 10 up to 1000	Recall
13	Multiplication square	25	Knowing and using number facts	Derive and recall multiplication facts for the 2-, 3-, 4-, 5- and 6-times tables	Recall
14	Bingo: multiplication	26	Knowing and using number facts	Derive and recall multiplication facts for the 2-, 3-, 4-, 5- and 6-times tables	Recall
15	Number line: checking calculations	27	Knowing and using number facts	Use knowledge of number operations and corresponding inverses to estimate and check calculations	Refine

Starter Number	Star Starter Title	Page No.	Strand	Learning objective as taken from the Primary Framework for Mathematics	Type of Starter
16	Maths Boggle: money problems	28	Knowing and using number facts	Use knowledge of number operations and corresponding inverses, including doubling and halving, to estimate and check calculations	Refine
17	Function machine: addition and subtraction	29	Calculating	Add or subtract mentally combinations of one-digit and two-digit numbers	Reason
18	Targets: addition and subtraction	30	Calculating	Add or subtract mentally combinations of one-digit and two-digit numbers	Rehearse
19	Maths Boggle: addition and subtraction	31	Calculating	Add or subtract mentally combinations of one-digit and two-digit numbers	Refine
20	Number line (ITP)	32	Calculating	Add or subtract mentally combinations of one-digit and two-digit numbers	Refine
21	Function machine: multiplication and division	33	Calculating	Use practical and informal written methods to multiply and divide two-digit numbers	Reason
22	Symmetry (ITP)	34	Understanding shape	Draw the reflection of a shape in a mirror line along one side	Refine
23	Maps and directions	35	Understanding shape	Read and record the vocabulary of position, direction and movement, using the four compass directions to describe movement about a grid	Rehearse
24	Find the cat: coordinates	36	Understanding shape	Read and record the vocabulary of position, direction and movement to describe movement about a grid	Reason
25	Sorting machine: shapes	37	Understanding shape	Identify right angles in 2D shapes	Reason
26	Weighing scales	38	Measuring	Know the relationship between kilograms and grams; choose and use appropriate units to estimate, measure and record measurements	Rehearse
27	Fixing points (ITP): using a ruler	39	Measuring	Read, to the nearest division and half division, scales that are numbered or partially numbered	Refresh
28	Measuring jug	40	Measuring	Read, to the nearest division and half division, scales that are numbered or partially numbered	Read
29	Clocks: reading a digital clock	41	Measuring	Read the time on a 12-hour digital clock	Read
30	Pictograms: favourite pets	42	Handling data	Collecting, organising and representing data; use pictograms to represent results and illustrate observations	Refine

Planning for the six Rs of oral and mental work

Oral and mental activity – six Rs	Using and applying mathematics	Counting and understanding number	Knowing and using number facts	Calculating	Understanding shape	Measuring	Handling data
Rehearse				• Targets: addition and subtraction	• Maps and directions	• Weighing scales	
Recall		• Pirates: adding multiples of 10	• Bingo: multiples of 2, 5 and 10 • Multiplication square • Bingo: multiplication				
Refresh		• Number square: counting forwards and backwards • Sorting machine: numbers				• Fixing points (ITP): using a ruler	
Refine	• Grouping (ITP) • Shopping: using money	• Fractions (ITP)	• Number spinners (ITP) • Number line: checking calculations • Maths Boggle: money problems	• Maths Boggle: addition and subtraction • Number line (ITP)	• Symmetry (ITP)		• Pictograms: favourite pets
Read						• Measuring jug • Clocks: reading a digital clock	
Reason		• Twenty cards (ITP): largest and smallest numbers • Bricks: ordering whole numbers • Dominoes: fractions of shapes	• Number sentence builder: find the missing number	• Function machine: addition and subtraction • Function machine: multiplication and division	• Find the cat: coordinates • Sorting machine: shapes		

Grouping (ITP)

Strand

Using and applying mathematics

Learning objective

Solve one-step and two-step problems involving numbers

Type of starter

Refine

Whiteboard tools

● 'Number of counters' button: press the arrows above and below the number to change the number of counters that are placed on the screen. This will also change the calculation below the number line. Press the number itself for the counters to appear.
● Press the arrows of the 'grouping' button to change the total number in each group; this will also update the calculation below the number line.
● Press the 'counter selector' button to change the type of counters. There are four types from which to choose.
● Press '=' to reveal the answer.

What to do

Use this ITP to help children to refine their skills in grouping numbers in order to investigate if they have a remainder. First, select how many counters you want on the screen (up to a maximum of 30), then press the number for the counters to appear on the screen. Then select how you would like these counters to be grouped. Press the counters to group them. Once they have all been grouped, use the number line to work out how many groups have been made, and whether there is any remainder. Press = to check the answers.

Differentiation

Less confident: use number tracks or multiplication grids to support children in grouping the numbers. Begin by using number sentences that have no remainders.
More confident: use numbers that have at least one remainder, such as 17 divided by 4. Challenge the children to find related multiplication facts.

Key questions

● What strategies can we use to accurately predict what the remainder might be? How can you prove it?
● Where might this strategy be useful in an everyday situation? (For example, sharing objects between a group of children.)

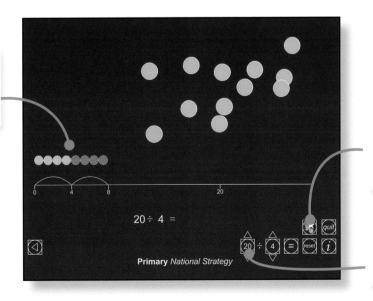

automatic grouping
Counters will automatically group themselves when pressed

counter selector
Press to change appearance of counters

number of counters
Press to change total number of counters used

Shopping: using money

Strand

Using and applying mathematics

Learning objective

Solve one-step problems involving money

Type of starter

Refine

Whiteboard tools
● Drag one item from the shop into the shopping basket.
● Press 'check-out' to take the item to the till.
● Drag the coins into the till to make the exact total to pay for the item.
● Press the 'sale' button to check if the amount paid is correct.
● Press 'back to shop' to start a new sale.

What to do

This activity is designed to familiarise children with all the coins up to £2.00, as well as helping them to understand and use the correct notation and pay for items using different combinations of coins. Initially, select the sticker album on screen 1 and ask children to come to the board to drag and drop different combinations of coins to pay for this item. Continue with items of higher value (for example, the toy car). Ask: *What items could you buy with only three coins?*

Differentiation

Less confident: initially select items with low price points to build the children's confidence, such as the sticker album (30p) or pen (55p).
More confident: ask the children to predict how many ways there are to pay for a particular item. Ask: *Which three items could you buy for £2.00? How much change would you have?*

Key questions
● *What coins would you use to pay for this item?*
● *How many coins do you need to pay for this item?*
● *Can you think of another way to pay for this item?*

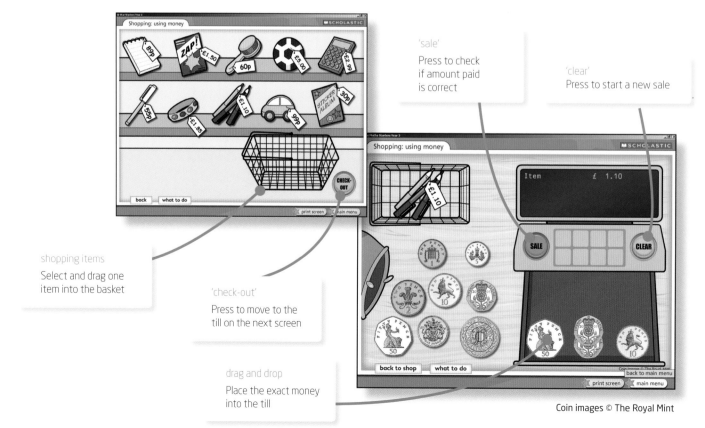

'sale'
Press to check if amount paid is correct

'clear'
Press to start a new sale

shopping items
Select and drag one item into the basket

'check-out'
Press to move to the till on the next screen

drag and drop
Place the exact money into the till

Coin images © The Royal Mint

Twenty cards (ITP): largest and smallest numbers

Strand

Counting and understanding number

Learning objective

Read, write and order whole numbers to at least 1000

Type of starter

Reason

Whiteboard tools

- Press the pack of cards with the blue outline and select 'random numbers' from the pop-up menu.
- Move the arrow to the right of 'how many cards' to 3.
- Move the arrow to the right of 'maximum number' to 9.
- Leave 'minimum number' at zero.
- Press 'go'.
- Press the central arrow on each card to drag and drop it into the middle of the screen.
- Press the red section of each card to reveal the number.
- Drag and drop each card to re-order them.
- Press the pack of cards with the white outline and select three cards to limit the number range.

What to do

The aim of this activity is to make the largest and smallest numbers possible using three single-digit cards. Press 'go' to select three cards from the pack at random. Ask the children to write the largest number possible using these three digits on their individual whiteboards and to hold up their answers. Check for any misconceptions, then re-arrange the cards on the board and ask individual children why they decided on a particular arrangement. Repeat the exercise using the same digits, but this time ask the children to make the smallest number possible. Encourage them to use the correct mathematical language. Extend the activity in subsequent sessions by asking the children to write the numbers on their boards in figures and in words.

Differentiation

Less confident: limit the number range using the 'make a stack' option.
More confident: extend the activity to four-digit numbers and ask the children to write the largest and smallest numbers in figures and in words.

Key questions

- *What is the largest/smallest number that can be made with these cards?*
- *How do you know this is the largest/smallest number that can be made?*

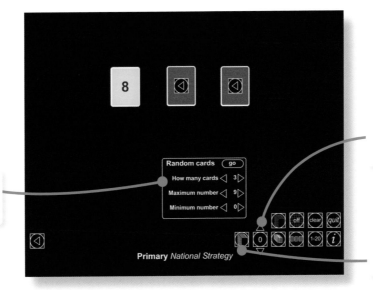

pack of cards
Press and select 'random numbers' from menu

'how many cards'
Move arrow to 3

'make a stack'
Press and select three cards to limit range of numbers

Bricks: ordering whole numbers

Strand

Counting and understanding number

Learning objective

Read, write and order whole numbers to at least 1000

Type of starter

Reason

Whiteboard tools
- Use the options to fix the first digit of the numbers, if required.
- Press 'go' to generate five bricks, each showing a number between 1 and 1000.
- Drag each brick into the gaps in the wall, with the smallest number in the lowest position, to complete the wall.
- If all five bricks are positioned correctly, a 'Well done' message appears. If any bricks are placed incorrectly, a 'Try again' message appears. Press 'ok' and the bricks move back to their starting position.
- Press 'go' again to select a new set of bricks.

What to do

Use this activity either to rehearse existing strategies for ordering numbers or to probe children's understanding. Ask the children initially to work as a whole class to decide the correct order of the bricks, but then to decide in pairs or individually by writing answers on their individual whiteboards. Position the bricks in the wall by dragging and dropping them (or ask individual children to place them). If the children place the bricks incorrectly, a message box appears inviting them to try again. Press 'ok' to move the bricks back to their starting order.

Differentiation

Less confident: support the children with individual number lines to help their ordering skills. Select the 'fix first digit' option to limit the range of numbers selected.
More confident: ask the children what would need to be added to the top brick to make, for example, 1000.

Key questions
- *What does each digit on the brick represent – hundreds, tens or ones?*
- *What would the new number be if 100, 200 or 300 were added to the number on the lowest brick? Or if 10, 20 or 30 were added to the number on the lowest brick?*

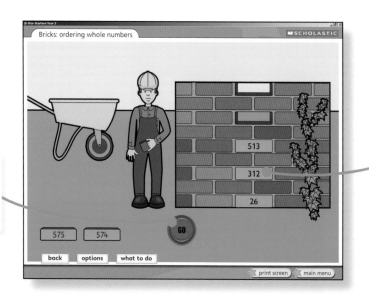

'go'
Press to generate a new set of bricks

bricks
Order by dragging into wall

Pirates: adding multiples of 10

Strand

Counting and understanding number

Learning objective

Count on from and back to zero in multiples of 10

Type of starter

Recall

Whiteboard tools

● Select options to change the pirate's base or start number from 0.
● Move the pirate up and down the mast in steps of 10 to answer the maths problem or number sentence shown on screen.
● Select options to generate different types of maths problem - select from 'up only' (addition) or 'up and down' options (addition and subtraction).
● Press 'new' to move the pirate back to the starting point and to generate a new question.
● Press 'notepad' and use the pen tool to show methods of working out the answers.
● Press 'answer' to reveal the completed number sentence.

What to do

Use this activity to encourage children to add or subtract mentally in steps of 10. Vary the base number and the type of maths problem through the options menu. Encourage the children to use their individual whiteboards so that they can work out the answers prior to any class discussion. Extend the activity beyond the initial question by asking, for example: *What happens if the pirate now moves down ten steps?* Illustrate this by dragging and dropping the pirate to the new position.

Differentiation

Less confident: use photocopiable page 43 to set some questions in which the pirate starts at zero to consolidate children's understanding.
More confident: ask the children to explain at the whiteboard, using the on-screen notepad, how they are calculating their answers.

Key questions

● *On which number will the pirate finish?*
● *How many places up or down would the pirate need to move in order to reach, for example, 50?*

'new'
Press to generate question

'options'
● Select 'up only' or 'up and down'
● Show or hide number sentence

pirate
Drag pirate up or down mast

Number square: counting forwards and backwards

Strand

Counting and understanding number

Learning objective

Count on, from and back to zero

Type of starter

Refresh

Whiteboard tools

● Press 'options' to select a number square from the following:

1 to 100 stepping in ones (select squares on side: 10; start number 1; step: 1)

2 to 200 stepping in twos (select squares on side: 10; start number: 2; step: 2)

5 to 500 stepping in fives (select squares on side: 10; start number: 5; step: 5)

● Press the yellow square to highlight numbers.
● Press the blue square to hide numbers.
● Press the white square to clear highlighted numbers and reveal hidden numbers.

What to do

The task for children is to count forwards or backwards in ones, twos or fives, depending on the choice of number square. Start by highlighting a number on the square - for example, 20. Ask the children to count forwards or backwards to this number. Then hide a number of squares adjacent to it and ask them to repeat this process up to the last hidden square. Clear this last hidden square to reveal the unknown number and check the children's counting accuracy.

Differentiation

Less confident: start with the 1-100 square, stepping in ones until the children understand the idea.
More confident: move on to squares using step sizes of 2 and 5, and select 2-200 and other number squares.

Key questions

Once some numbers are hidden, ask:

● *What is the number just before or just after the number showing?*
● *What is the number 3 before or 3 after the number showing?*

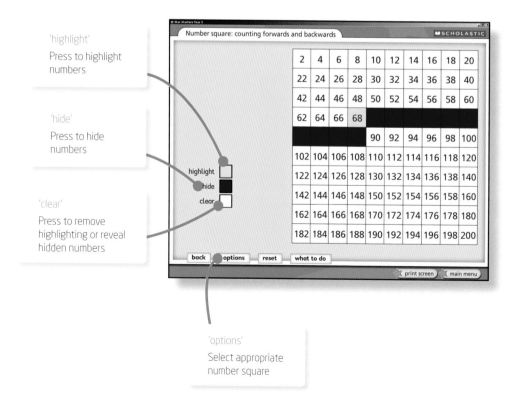

'highlight'
Press to highlight numbers

'hide'
Press to hide numbers

'clear'
Press to remove highlighting or reveal hidden numbers

'options'
Select appropriate number square

Sorting machine: numbers

Strand

Counting and understanding number

Learning objective

Round two-digit or three-digit numbers to the nearest 10 or 100

Type of starter

Refresh

Whiteboard tools

- Press 'go' to launch the first ball at the top of the screen into the sorting machine.
- Press one of the cogs of the sorting machine to decide whether it is a two-digit or three-digit number.
- Press one of the cogs on the second row of the sorting machine to decide how the selected number should be rounded.

What to do

Use this starter to identify the appropriate nearest whole number for both hundreds or tens units. Press 'go' to launch the first number. For each number produced by the computer, ask the children to vote whether it is a tens or hundreds number, and then which number it should be rounded to. Encourage the children to explain their selections. Make sure that they all understand that we round up when the number is half way between two tens (for example, 85 always rounds up to 90). Check the children's answers and discuss any misconceptions.

Differentiation

Less confident: support the children by giving them sets of place value cards.
More confident: ask the children to predict which bucket the number should go into when it first appears.

Key questions

- *What part of the number are we investigating in order to sort it accurately?*
- *How do you know which bucket the number should be sorted into?*
- *What if you had to sort 85 or 150? Would you round up or down?*

'go'
- Press to launch each number

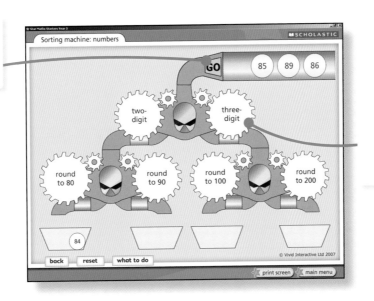

cogs
- Press to select route for number

Dominoes: fractions of shapes

Strand

Counting and understanding number

Learning objective

Identify and estimate fractions of shapes

Type of starter

Reason

Whiteboard tools

- Press 'new' to start a new game.
- Press 'miss a go' to take another domino from the pot.
- Drag and drop the dominoes into the playing space. Rotate them 90° by pressing the top right-hand corner.
- Press 'winner' if Player 1 or Player 2 has placed all of the dominoes and it is agreed that the last domino was placed correctly.

What to do

The aim of this game is to help children visualise the fractions of a shape (a circle) by matching the fraction notation to shapes. Make sure that they understand that the shaded part of the circle represents the fraction they are trying to identify. The game is played in the same way as standard dominoes, with two players or groups playing against each other. Each player (or team) is dealt five dominoes, with the others left in a central pot. A starter domino is selected by the computer to begin the game, and the players then take turns to play. If you are placing the dominoes for the children, make sure at all times that the children use correct mathematical vocabulary to describe the shapes they wish to place.

If a player is unable to place a domino, they should press 'miss a go' and take one from the central pot. The game continues until a player places all of their dominoes, and is declared the winner, or there are no more dominoes left in the pot. If a stalemate situation occurs – in which neither player is able to put down a domino and the pot is empty – the player with fewer remaining dominoes is the winner. Encourage the children to describe the shapes on the domino before placing each domino.

Differentiation

Less confident: let the children use talk partners to discuss moves, which will help to boost confidence and affirm their decisions. Ask them to draw the fractions of the circle on paper to check their answers.
More confident: play 'beat the teacher', in which children pit themselves against an adult in the classroom.

Key questions

- *Which fractions were easy to spot? Which were difficult? Why?*
- *Can you find a shape fraction that shows one-third? One-tenth?*

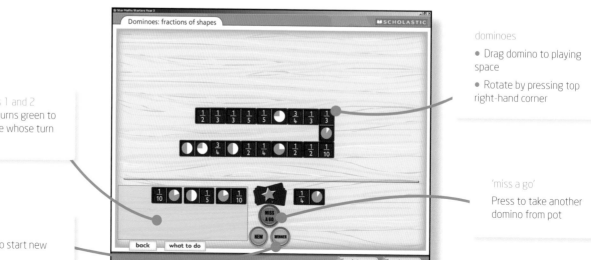

players 1 and 2
Panel turns green to indicate whose turn it is

'new'
Press to start new game

dominoes
- Drag domino to playing space
- Rotate by pressing top right-hand corner

'miss a go'
Press to take another domino from pot

Fractions (ITP)

Strand

Counting and understanding number

Learning objective

Use diagrams to compare fractions and establish equivalents

Type of starter

Refine

Whiteboard tools

- Press the small green and yellow bar to display more fraction bars, up to a maximum of five.
- Press the arrows next to the fraction bar to increase or decrease the denominator and show the fraction chosen.
- Press the 'fdpr' button to show/hide different fractions that you make.
- Press 'reset' to clear all but the lowest green bar.

What to do

This is a flexible tool that enables you to create a range of fractions and equivalent fractions. Start with four fraction bars to show a whole, halves, quarters and eighths by changing the denominators using the arrow keys on the right of the screen. Ask the children how many halves make a whole, then how many quarters make a whole and so on. Next, ask them how many quarters make a half and how many eighths make a half. Show them how these fractions are equivalent.

Any number of different equivalent fractions may be built up in this way to help the children see and understand equivalence. It is a good idea to leave the lowest fraction as a whole and then use the next bar up to demonstrate the most basic fraction you are investigating.

Differentiation

Less confident: start slowly, using only halves and quarters to make sure that the children understand the principles.
More confident: ask the children for their suggestions for fractions to build and ask what the equivalents might be.

Key questions

- *If the denominator is 3, how many sections will the bar be divided into?*
- *How many different ways do you think you could make a half?*

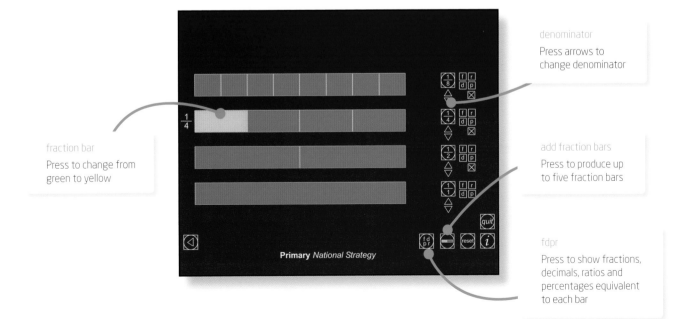

denominator
Press arrows to change denominator

fraction bar
Press to change from green to yellow

add fraction bars
Press to produce up to five fraction bars

fdpr
Press to show fractions, decimals, ratios and percentages equivalent to each bar

Number sentence builder: find the missing number

Strand

Knowing and using number facts

Whiteboard tools

- Move cards and symbols onto the line to build a number sentence.
- Drag and drop numbers and symbols within a line to re-order them.
- Drag cards off the line to remove them.
- Press 'reset' to clear the cards and symbols from each line.

Learning objective

Derive and recall all addition and subtraction facts for each number to 20

What to do

The aim of the activity is to find the missing number (or numbers) in a number sentence so that both sides of the equals sign balance. Use addition and subtraction methods only, such as 3 + △ = 20. Numbers of any size can be selected as two or more digits, when selected consecutively, snap together to form a longer number. For Year 3 use numbers up to 20 in order to tease out children's understanding of number facts rather than asking them to complete any complex calculations.

Type of starter

Reason

Differentiation

Less confident: start with straightforward sentences such as 3 + 10 = △ until the children become familiar with the idea that the symbol represents a number.
More confident: ask the children to make up their own number sentences to challenge the whole class.

Key questions

- *What is the missing number?*
- *Can you think of any number sentences using addition and subtraction that don't work?* (For example, 24 + 7 + △ = 25.)

drag and drop
Move cards and symbols onto lines

'reset'
Press to clear all the lines of numbers

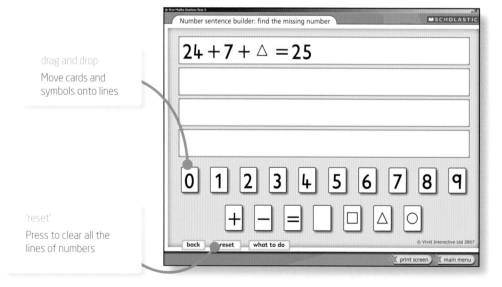

Number spinners (ITP)

Strand

Knowing and using number facts

Learning objective

Derive and recall all addition and subtraction facts to 20

Type of starter

Refine

Whiteboard tools

- Press the arrows on the left-hand button to increase or decrease the maximum number allowed on the spinners.
- Press the centre button to increase or decrease the number of sides on the spinners. There are three- to six-sided spinners available.
- Press the right-hand button to increase or decrease the number of spinners. There are 1-3 spinners available. Select the number to create the spinners.
- Press the yellow dot in the centre of the spinner to spin the spinner.

What to do

Use this ITP to rehearse number facts to 20. Initially select two four-sided spinners with a number range to 15. Spin each spinner and ask the children to add or subtract the two numbers generated as appropriate, then to write the answers on their individual whiteboards. Ask them to hold up the whiteboards and check for any incorrect answers. Extend the activity by selecting spinners with five or six sides and by increasing the number range to 20.

You can also use this ITP to create incomplete number sentences to rehearse number bonds up to 20. For example, if 9 and 7 are spun, the numbers could be used to create two sentences: 9 + 7 + ? = 20, or 20 - 9 - 7 = ?. Highlight the fact that the subtraction is the inverse of the addition.

Differentiation

Less confident: reduce the number range to 1-5 and modify further by pressing a number generated by the spinner. This will increase the number by 1 to allow further practice and support for less confident learners.

More confident: select a higher number range (numbers to 50, for example) and three spinners.

Key questions

- *What is the missing number?*
- *How could you write this as an addition or subtraction number sentence?*

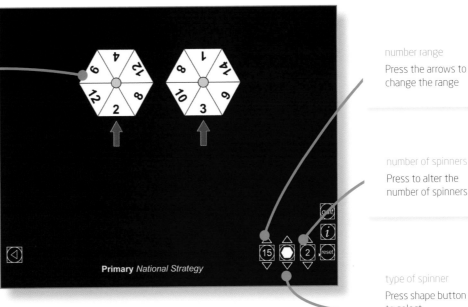

spinners
- Press the yellow dot to spin
- Press an individual number to increase it by 1

number range
Press the arrows to change the range

number of spinners
Press to alter the number of spinners

type of spinner
Press shape button to select

Primary National Strategy

Bingo: multiples of 2, 5 and 10

Strand

Knowing and using number facts

Learning objective

Recognise multiples of 2, 5 or 10 up to 1000

Type of starter

Recall

Whiteboard tools

- Print bingo cards from the opening screen.
- Set the timer in 'options' to adjust the time between bingo calls (5–20 seconds).
- Press the 'start' button to start a new game.
- Press 'check grid' to check the answers if someone calls *House*.
- Press 'play on' or 'winner' after checking a player's grid.

What to do

This activity is designed to use number facts and encourage quick recall of multiples of two-digit numbers by 2, 5 or 10. Ask the children to play individually or organise them into pairs; provide each child or pair with a bingo card (these can be printed from the opening screen or prepared using the bingo card template on page 44).

Each ball offers a different multiplication sentence. Explain to the children that if the answer appears on their bingo grid, they should mark it off. If the children are new to the game allow for a longer amount of time between bingo calls. If a child calls *House* (or other similar winning call), press the 'check grid' button to pause the game and call up the checking grid, which includes all of the number sentences that have been called. If they are correct, press the 'winner' button to hear an appropriate fanfare. If they are not correct, press 'play on' to continue the game.

Differentiation

Less confident: if time is limited, ask children to call *House* once they have correctly ticked off four numbers (or however many time allows).
More confident: reduce the time between bingo calls or prepare bingo cards with more numbers for the children to find.

Key questions

- *Were any of the answers the same? If so, why?*
- *Where do we come across numbers that are multiplied by 10 or 100?* (Consider the metric system, for example, where 23cm is equal to 230mm.)

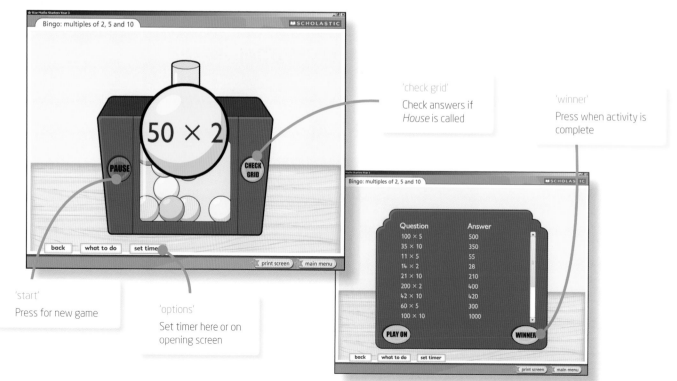

Multiplication square

Strand

Knowing and using number facts

Learning objective

Derive and recall multiplication facts for the 2-, 3-, 4-, 5- and 6-times tables

Type of starter

Recall

Whiteboard tools
- Select 'clear' or 'highlight' and then press any square to reveal the number beneath.
- Select 'hide' to hide any numbers that you have revealed.

What to do
Prepare the multiplication square before the lesson: select one number on the multiplication grid and leave all of the remaining numbers hidden. Challenge the children to find all the other numbers in the multiplication square up to 60. The key to this activity is in the selection of questions as this will draw out the children's understanding of the times tables to up to 6 × 10. After each question, press the square to reveal the number and confirm the children's answers. Ask for more than one answer before revealing the missing number.

Differentiation
Less confident: ask the children to contribute answers later in the session when some patterns of answers have already been established.
More confident: ask the children to tell you the whole row or column once one number in the row or column has been established.

Key questions
- *What are the numbers above and below the number showing?*
- *What are the numbers to the right and left of the number showing?*
- *Who can tell me the numbers in a whole row or column?*

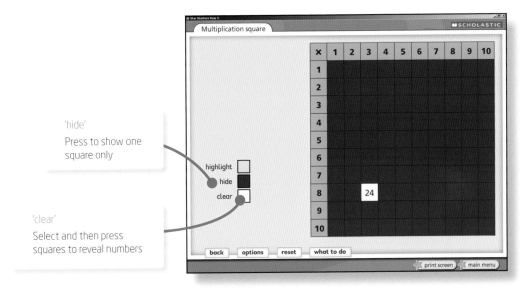

'hide'
Press to show one square only

'clear'
Select and then press squares to reveal numbers

Bingo: multiplication

Strand

Knowing and using number facts

Learning objective

Derive and recall multiplication facts for the 2-, 3-, 4-, 5-, 6- and 10-times tables

Type of starter

Recall

Whiteboard tools

- Set the timer to adjust the time between bingo calls (5–20 seconds).
- Press the 'start' button to start a new game.
- Press 'check grid' to check the answers to the number sentences if someone calls *House*.
- Press 'winner' when a player has successfully completed their grid; press 'play on' if the grid has not been completed correctly.

What to do

This activity is designed to rehearse number facts and encourage quick recall of multiples up to 10 × 10 and their corresponding division facts against a time limit. Ask the children to play individually or organise them into pairs; provide each child or pair with a bingo card (these can be printed from the opening screen or prepared using the bingo card template on page 44).

Each ball offers a different number sentence. Explain to the children that if the answer to the number sentence appears on their bingo grids, they should mark it off. If they are new to the game, allow for more time between bingo calls. If a child calls *House* (or other similar winning call), press the 'check grid' button to pause the game and call up the checking grid, which includes all of the number sentences that have been called. If they are correct, press the 'winner' button to hear an appropriate fanfare. If they are not correct, press 'play on' to continue the game.

Differentiation

Less confident: if time is limited, ask the children to call *House* once they have correctly ticked off four numbers (or however many you feel time allows).
More confident: reduce the time between bingo calls or prepare bingo cards with more numbers for the children to find.

Key questions

- *Were any of the answers the same? If so, what patterns can be seen in the numbers?*
- *If you use a multiplication grid, are you able to find the product more quickly?*

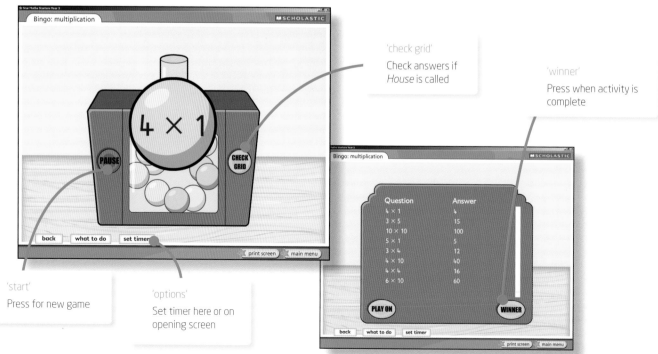

'check grid'
Check answers if *House* is called

'winner'
Press when activity is complete

'start'
Press for new game

'options'
Set timer here or on opening screen

Number line: checking calculations

Strand

Knowing and using number facts

Learning objective

Use knowledge of number operations and corresponding inverses to estimate and check calculations

Type of starter

Refine

Whiteboard tools

● Drag and drop the red pointer across the number line to position it.
● Drag and drop the green pointer across the number line to position it.
● Use the 'options' button to hide the marker numbers.

What to do

Move the red marker to a number between 1 and 100 and ask children to read it. Ask them to count on in ones to the nearest 10 above. Then ask them to count on in tens up to 100. Make a note of the number pair found as a number sentence: for example, if the red marker position is 34 then 34 + 66 = 100. Next, position the red marker at 66 and ask children what they think they will need to add to make 100. Count on again if necessary. Write the new number sentence underneath the first. Find number sentences for other number pairs totalling 100. Ask the children, in pairs, to check calculations using the corresponding subtraction sentences.

Differentiation

Less confident: position the green pointer at a lower number, such as 30, then position the red pointer to a number less than 30. Ask the children to count on from the number on the red pointer up to 30.
More confident: ask the children to find the difference between the number on the red marker and 100 without counting on, and to write down both number sentences without any further calculations.

Key questions

● *If two numbers added together make 100, does it make any difference which way round they are written in a number sentence?*
● *If 1, 2 or 3 is added to the red number marker, what number would now be needed to make 100?*

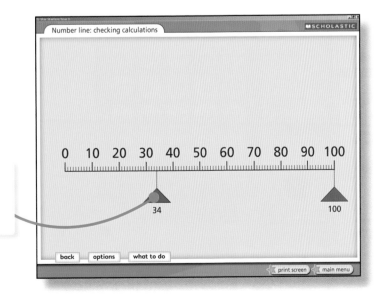

markers
Drag to the desired position on the number line

Targets: addition and subtraction

Whiteboard tools

- Press 'go' to generate four number cards and a target number.
- In the 'options' menu, select 'randomly generated' for the program to generate a target number, or 'entered by teacher' to manually insert a target number.
- Use the 'notepad' to show calculations.
- Use the 'pen' tool or write on the notepad. Press 'start again' to delete any text, or use the 'eraser' tool.
- Press 'winner' if a child completes the activity successfully to see the winner's fanfare.

What to do

The aim of this activity is for the children to use and rehearse known number facts to find a target number. Press 'go' to display four cards and a target number. Challenge the children to make the target number by adding and subtracting the numbers on the four cards. Alternatively, you can type your own target number by selecting 'entered by teacher' in the 'options' menu. Encourage the children to use known strategies in order to write number sentences that match (or nearly match) the target number as not every target will be achievable.

Invite individual children to write their calculations on the 'notepad' (using the pen tool) and review the process. At this point, encourage the rest of the class to challenge the process - or suggest an alternative method.

Differentiation

Less confident: if necessary, provide the children with number squares or individual number lines. For additional support provide each child with a copy of the photocopiable 'Targets' sheet on page 45.
More confident: extend the activity by asking the children to use simple multiplication and division facts. Also, set some target numbers above 100 using the 'entered by teacher' mode.

Key questions

- What strategies are the most efficient? How do they help you to 'hit the target'?
- What other ways are there to make (or get close to) the target number?

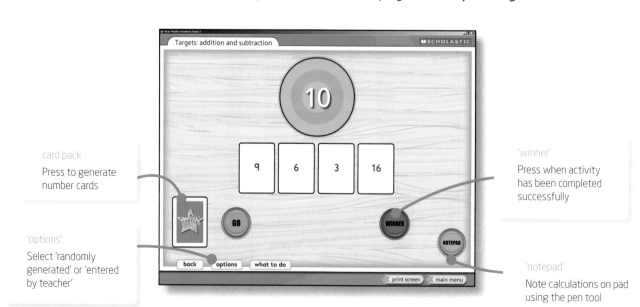

card pack
Press to generate number cards

'options'
Select 'randomly generated' or 'entered by teacher'

'winner'
Press when activity has been completed successfully

'notepad'
Note calculations on pad using the pen tool

Maths Boggle: addition and subtraction

Strand

Calculating

Learning objective

Add or subtract mentally combinations of one-digit and two-digit numbers

Type of starter

Refine

Whiteboard tools

- Press 'new' to rattle the dice and start a new game.
- Highlight each dice by pressing it once (to remove the highlight, press again).
- Change the target by selecting a new question from the 'options' menu at the foot of the screen.
- Use the 'notepad' to show calculations.

What to do

The aim of this activity is to use mental methods of addition and, where this is well established, to begin to use more sophisticated strategies to estimate which number strings (columns or rows) are most likely to contain the target answers.

Start by selecting a question from the 'options' menu at the foot of the screen (or, should you wish, by setting your own question). The dice are 'rattled' to reveal a random selection of numbers. In pairs or individually, the children find the answer to each question by calculating using the numbers on the screen. Answers can be checked by highlighting individual rows or columns. Challenge children to come to the board to show their calculations using the on-screen notepad.

Differentiation

Less confident: limit the range of questions, focusing on pairs of numbers.
More confident: ask the children extension questions such as: *Which column or row would have the largest/smallest totals?*

Key questions

- *How can we estimate which rows or columns have the largest totals?*
- *How can we quickly check answers?* (For example, doubling common numbers or finding multiples.)

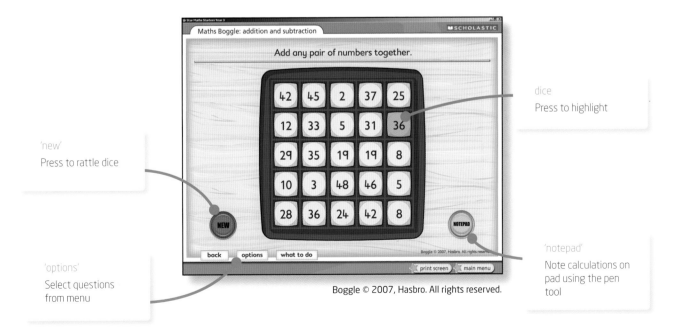

'new'
Press to rattle dice

'options'
Select questions from menu

dice
Press to highlight

'notepad'
Note calculations on pad using the pen tool

Boggle © 2007, Hasbro. All rights reserved.

Number line (ITP)

Strand

Calculating

Learning objective

Add or subtract mentally combinations of one-digit and two-digit numbers

Type of starter

Refine

Whiteboard tools

● Drag the discs below the number line to set the start and end numbers.
● A subtraction sentence can be made by dragging the left button over the right button. The sentence will automatically show a subtraction at this point. To return to an addition question, drag the buttons in the opposite way.
● Press the first circle to fix it; when fixed, the circle is yellow. Press the arrows next to the 'min' and 'max' buttons to change the range of the number line. There is a range of −30 to 500.
● The numbers above the number line can be shown or hidden.

What to do

The aim of this activity is to refine the skills needed when adding or subtracting pairs of numbers, using a number line as a support. Press the arrows either side of the 'min' and 'max' buttons on the ITP to set the range of the number line. Drag the first disc at the beginning of the line to 70 and the second disc to 30. Press the 'first number' and 'second number' buttons to add labels to the number line and ask the children to predict the difference. Move the right-hand disc a few times to allow the children to practise and then challenge them to find the difference if the numbers on the number line are missing. Press the 'numbers on the scale' button to hide the numbers and repeat the activity on a blank number line.

Differentiation

Less confident: support children by giving them their own number line so that they can reflect the numbers on the board on their individual whiteboards at their desks.
More confident: select higher maximum numbers (up to 500), or hide the numbers on the number line and the number sentence, and ask the children to complete a range of questions.

Key questions

● *How can you use the number line to estimate the answer?*
● *What happens when we change the second number?*

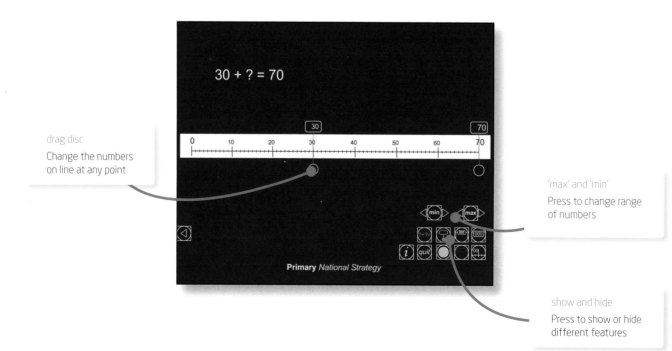

drag disc
Change the numbers on line at any point

'max' and 'min'
Press to change range of numbers

show and hide
Press to show or hide different features

Function machine: multiplication and division

Strand

Calculating

Learning objective

Use practical and informal written methods to multiply and divide two-digit numbers

Type of starter

Reason

Whiteboard tools

- Use the 'options' menu to set the 'machine mode'. Select from 'manual' or 'random' options.
- Select 'manual' to prepare your own number sentences, or 'random' to produce a computer-generated number sentence.
- A keypad pops up automatically when you press on a window to enter a number.
- Press the 'history' button to view a list of the number sentences completed during the lesson.

What to do

The aim of this game is to complete a number sentence by finding the hidden number or function. Either the teacher or the computer can generate these numbers using either the 'manual' or 'random' mode options.

Manual mode: prepare two of the three boxes and challenge the children to work out what number or function is in the third box. (If you are using this on a whiteboard you might wish to switch between your PC screen and the whiteboard to prepare the number sentences.) Prepare some number sentences involving the multiplication or division of two-digit numbers, such as 21 ÷ 7, 50 ÷ 4, 13 × 3, 11 × 5 and so on. Input the numbers into the function machine using the keypad. Press 'go' to check the answers.

Random mode: the computer selects a number sentence, but hides the input, output and function windows on the machine. Decide which element to reveal first and press that window to open it. After one other element has been revealed, ask the children to write down and then display the missing number or function. Check their answers and then press 'go' to check the answer on the machine.

Differentiation

Less confident: hide the input number and function performed to show just the final number and function number. Ask the class to think of a number sentence that would make the given answer.

More confident: patterns and sequences can be demonstrated using the machine. For example, multiplying by 5 can be modelled by keeping the function to ×5.

Key questions

- *How are we able to find missing parts of the sentence?*
- *How much of the sentence needs to be revealed before we can complete it?*

'new'
- Press to start again in 'manual' mode
- Press for number sentence in 'random' mode

'options'
- Select 'manual mode' to enter your own numbers
- Select 'random mode' for computer-generated numbers, initially hidden

windows
- Type numbers and functions in 'manual' mode
- Press to open in 'random' mode

'history'
Use to track completed number sentences

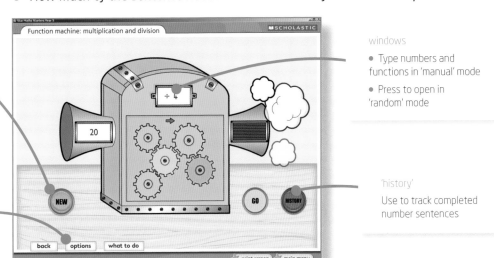

Symmetry (ITP)

Strand

Understanding shape

Learning objective

Draw the reflection of a shape in a mirror line along one side

Type of starter

Refine

Whiteboard tools

- Press the squares on the grid to create a shape.
- Press the 'reflection' button to show or hide the reflection.
- Press the 'shape direction' arrows to move a shape. If the 'reflection' button has been selected the reflection will move at the same time.
- Press the 'mirror' button to change the line of reflection.

What to do

The aim of this activity is for children to refine their ability to draw the reflection of a shape. The ITP can also be used to check predictions.

When the program is launched a blank grid is displayed. Create a shape by pressing the squares on the grid (for example, the shape in the image below uses only four squares). Any highlighted squares appear in yellow; pressing each square again will clear it. Once created, a shape can be moved around the screen by using the 'shape direction' arrows. By default the mirror is placed vertically but as the children get used to the program the line can be changed by pressing the 'mirror' button.

Differentiation

Less confident: pressing either of the four direction arrows can move any shape. Ask the children to predict where the shape will be once it has been moved an agreed number of squares.

More confident: encourage more confident children to change the angle of the mirror and predict how the reflected shape will move.

Key questions

- *Which squares will the reflection fall into when the shape is moved?*
- *What effect does changing the mirror angle have on the reflection?*

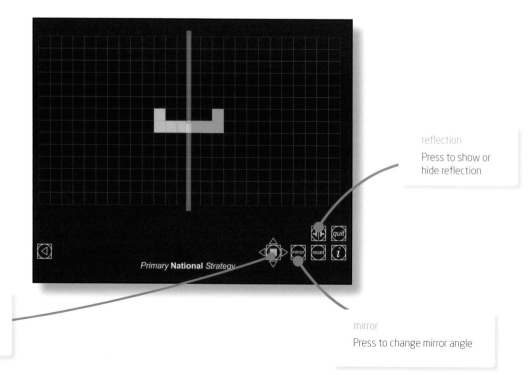

reflection
Press to show or hide reflection

shape direction
Press arrows to move shape

mirror
Press to change mirror angle

Maps and directions

Strand

Understanding shape

Learning objective

Read and record the vocabulary of position, direction and movement, using the four compass directions to describe movement about a grid

Type of starter

Rehearse

Whiteboard tools
● Drag and drop the direction and movement cards to prepare the route. Press 'move' to confirm the chosen route.
● Press 'show route' to display the directions selected so far.
● View the box at the top of the screen to identify which items the knight has collected along the route.

What to do
This activity develops children's use of the four compass points: north, east, south and west. From a given starting point, ask the children for directions to guide the knight through the castle, using the direction and movement cards on the screen as prompts. The knight has to collect his sword, shield and helmet before making his way to the exit. Drag and drop each card into place to build up the route. The order in which the cards are placed affects whether the knight turns first or moves first. Press 'move' after each selection.

More than one route is available and some routes include barriers – so the quickest route is not necessarily the best. The position of the items and barriers changes randomly with each game. Press the 'show route' button to check the knight's progress. Encourage the children to check and challenge the route at this stage and start again if necessary. Make sure at all times that they use the correct vocabulary when selecting an instruction from the screen.

Differentiation
Less confident: ask the whole class to show you with their arms the directions for north, south, east and west before starting the activity to check understanding. Provide copies of photocopiable page 46 for additional support.
More confident: ask the children for alternative ways of giving the same direction (for example: *moving through a right-angle towards the left*).

Key questions
● *What is the most direct way to move from this point to that point?*
● *Are there any alternative ways of moving from this point to that point?*

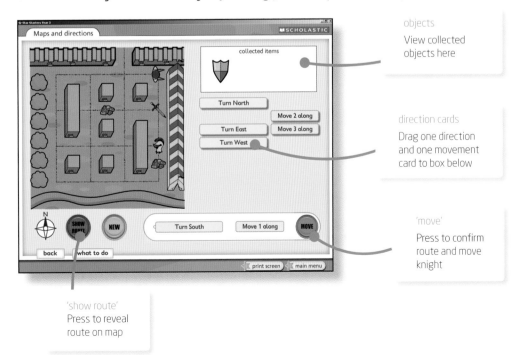

objects
View collected objects here

direction cards
Drag one direction and one movement card to box below

'move'
Press to confirm route and move knight

'show route'
Press to reveal route on map

Find the cat: coordinates

Strand

Understanding shape

Learning objective

Read and record the vocabulary of position, direction and movement to describe movement about a grid

Type of starter

Reason

Whiteboard tools

- Type in the horizontal coordinate, then the vertical coordinate, to select a square.
- Press 'check' to confirm the choice.
- Continue to type in coordinates until the cat has been found.
- Press 'new' to start a new game with the cat in a different position.

What to do

The aim of the activity is to find a square on a 4 × 4 grid in which a cat is hiding. Tell the children that they should use coordinates to identify each square. If necessary, explain that they should give the horizontal coordinate before the vertical coordinate (for example, 2, 3 is two squares along and three squares up). The position of the cat is randomly selected each time. After each selection, a square will be revealed. To narrow the selections down and prevent the activity becoming a guessing game, a message appears after each selection: 'vertical correct' or 'horizontal correct'. If neither the row nor the column is correct then 'have another go!' appears. When the cat has been found, the garden scene is completed and a 'miaow' is sounded. Children should respond positively to this activity and will start to develop logical processes to find the cat in the fewest selections.

Differentiation

Less confident: ask the children for the horizontal and vertical numbers of several squares to make sure that they understand the process involved.
More confident: ask the children about possible strategies to find the cat in the fewest attempts.

Key questions

- *How can you describe, for example, the square in the top left-hand corner using horizontal and vertical numbers?*
- *Which possible squares could the cat be in, now that we know the row or column it is hiding in?*

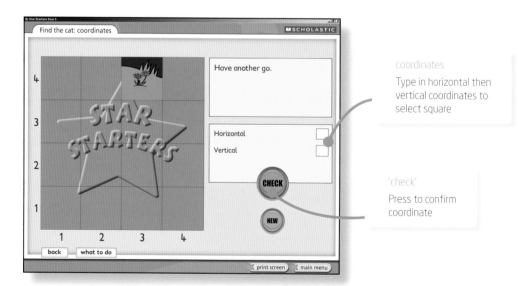

Sorting machine: shapes

Whiteboard tools

- Press 'go' to launch each shape in the sorting machine.
- Press one of the cogs of the sorting machine to decide whether or not the marked angle in the shape is a right angle or not.

What to do

The aim of this starter is to identify whether a highlighted angle in a shape is or isn't a right angle. Press 'go' to launch the first shape. For each shape selected, ask the children to vote on whether the identified angle is a right angle, then ask individual children to give reasons for their selections using appropriate mathematical language. Press the appropriate cog to sort the shape into the chosen bucket. Discuss any misconceptions with the children.

Differentiation

Less confident: give the children a set of prepared 2D shapes to test whether an angle is a right angle.
More confident: ask the children at the start of the lesson to predict how many angles are right angles and check the predictions at the end. Also, ask if any of the shapes have more than one right angle.

Key questions

- *How can we tell if a shape has or has not got a right angle?*
- *Where can we find right angles around the classroom?* (Books, desks, whiteboard and so on.)
- *Can we draw any conclusions as to which shapes have right angles and which shapes do not?*

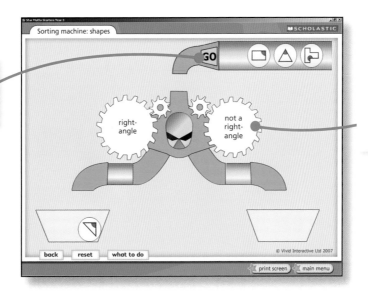

'go'
Press to launch each shape

cogs
Press to select bucket for shape to be sorted into

Weighing scales

Strand

Measuring

Learning objective

Know the relationship between kilograms and grams; choose and use appropriate units to estimate, measure and record measurements

Type of starter

Rehearse

Whiteboard tools

● Drag and drop items on the left of the screen into the pan.
● Turn the digital readout on or off on the analogue scale, as required.
● Select 'options' to make changes to the face of the scales, such as removing the numbers, or altering the number of subdivisions.

What to do

The aim of this starter is to rehearse estimating the weight of an item in grams and kilograms. Ask a child to choose an item from the left of the screen then ask the class to vote on whether they think it will it weigh more or less than a kilogram. Discuss the difference between a kilogram and a gram and which objects they would estimate would weigh more or less than a kilogram. Test the children's predictions by dragging and dropping the items onto the scales. Ask them also to predict how many of a particular item would be needed to make 1kg (for example, two cartons of milk weigh 1kg - 500g × 2).

Differentiation

Less confident: ask the children to read the weight in kilograms and grams. Turn on the digital readout to support the children with this activity.
More confident: ask the children to estimate how many of the smaller items (for example, apples) would weigh the equivalent of a kilogram. Ask them to make up different combinations of items to make 1kg exactly (for example, four bags of crisps, one loaf of bread and one bunch of grapes weigh exactly 1kg).

Key questions

● *How much does this item weigh in grams? What does it weigh in kilos?*
● *We know that ___ weighs ___, so how many would be needed to make a kilogram?*

products
Drag food items onto the scales

'options'
Set maximum weight, subdivision, and whether or not to show numbers on analogue scale

scales pan
Add or remove items by dragging them on or off pan

digital readout
Can be turned on or off, on analogue scales, to reveal exact weight of items

Fixing points (ITP): using a ruler

Strand

Measuring

Learning objective

Read, to the nearest division and half division, scales that are numbered or partially numbered

Type of starter

Refresh

Whiteboard tools

- Press 'grid size' to change the size of the grid.
- Press any two points of the grid to create a line between them.
- Drag one end of the line to another grid point to draw lines of different lengths more quickly.
- Press the 'ruler' button to show or hide the ruler at the bottom of the screen.
- Use the blue circle at one end of the ruler to rotate it and drag it into the correct position for measurement.
- Press 'reset' to clear the grid.

What to do

Use this activity to demonstrate how to read scales and how to use a ruler. This ITP enables lines of different lengths to be drawn very quickly and accurately. Invite different children to come to the board to select different lines and ask the class questions such as: *What length is shown? What length would be midway between the two points? Write the length to the nearest half centimetre.*

Ask the children to write the lengths on their individual whiteboards using the correct decimal notation.

Differentiation

Less confident: ask the children to read the scale to the nearest centimetre initially.
More confident: encourage the children to give a more accurate reading, such as to the nearest millimetre.

Key questions

- *Where should the ruler be positioned to take the measurement: at the end of the ruler - at 0cm or 1cm?*
- *How do you write down the measurement that is half way between 3cm and 4cm?*

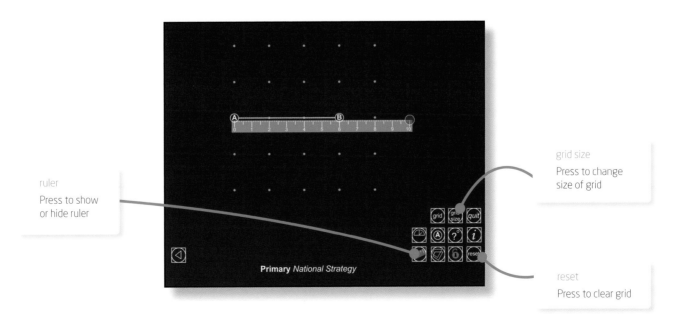

ruler
Press to show or hide ruler

grid size
Press to change size of grid

reset
Press to clear grid

Measuring jug

Strand

Measuring

Learning objective

Read, to the nearest division and half division, scales that are numbered or partially numbered

Type of starter

Read

Whiteboard tools

● Press the 'options' button to set the following: set 'scale' to 100ml; set 'subdivisions' to 2 (to give 5ml intervals on the jug); set 'fill steps' to manual.
● Press 'in' to fill the jug, and press it again to stop filling.
● Press 'out' to empty the jug, and press it again to stop emptying.
● Press 'reset' to start again.

What to do

Ask the children to read aloud together the marked divisions. Then question them about the unmarked divisions and ask them to read these aloud. Fill the jug part-way up to a certain level and then question the children about the reading. Fill and empty the jug to find the difference between the two levels. Set some capacity problems for the children and ask them to write the answers on their individual whiteboards. (For example: *A bottle of water holds 350ml. Sasha drinks 100ml. How much water is left in the bottle?*) Check the answers using the measuring jug - though you will need to change the scale.

Differentiation

Less confident: in the early stages ask the children to read the scale to the nearest 10ml or 100ml mark.
More confident: challenge the children by asking if they can suggest a more accurate answer than reading to the nearest 5ml. Use the 'options' menu to change the scale to 5 subdivisions.

Key questions

● *A large spoon holds 20ml. How many spoons can be filled from the full container?* (Set container to 100ml scale.)
● *A newborn baby has a bottle containing 100ml of milk. If she only drinks 80ml, how much is left?*

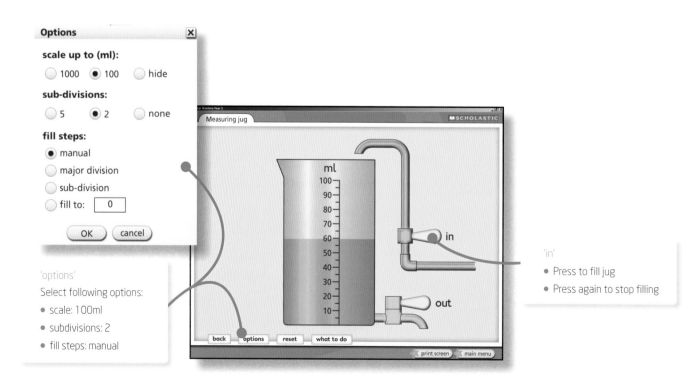

Options ☒

scale up to (ml):
○ 1000 ● 100 ○ hide

sub-divisions:
○ 5 ● 2 ○ none

fill steps:
● manual
○ major division
○ sub-division
○ fill to: 0

(OK) (cancel)

'options'
Select following options:
● scale: 100ml
● subdivisions: 2
● fill steps: manual

Measuring jug · SCHOLASTIC

ml
100
90
80
70
60 — in
50
40
30
20
10 — out

back · options · reset · what to do

print screen · main menu

'in'
● Press to fill jug
● Press again to stop filling

Clocks: reading a digital clock

Strand

Measuring

Learning objective

Read the time on a 12-hour digital clock

Type of starter

Read

Whiteboard tools

- Press the 'randomise' button in order to generate random times on the digital clock.
- Press the '+' and '−' buttons on the clock to set the time.
- Press 'hour mode' to change between a 12-hour and 24-hour clock.

What to do

The aim of this starter is to practise reading the time on a digital clock. Select a time using the 'randomise' button and ask the children to whisper the time to their partners. Then, on cue, ask them to call out the time. At this point challenge any misconceptions or inaccuracies before randomly selecting another time. Speed up the process as the children grow more confident in reading the time accurately. With a time such as 6.35, ask the children to call out both options, ie 35 minutes past 6 or 25 minutes to 7.

Differentiation

Less confident: set some specific times to the hour or half hour as a warm-up to the 'randomised' selection (for example, 12.00, 8.30 and so on).
More confident: change to a 24-hour clock and explain that a day can be measured in two 12-hour blocks. Challenge the children to read 24-hour times using am or pm conventions.

Key questions

- *What do the two figures represent on the clock face?*
- *What happens when the minute hand reaches 59?*
- *What is the difference between am and pm (antemeridian and post meridian), and why does it matter?*

minute/hour time set
Press to change time manually

'randomise'
Press to allow program to generate random times

Pictograms: favourite pets

Strand

Handling data

Learning objective

Collecting, organising and representing data; use pictograms to represent results and illustrate observations

Type of starter

Refine

Whiteboard tools

- Type in the title of the chart and labels for the categories.
- Press the arrow next to the red square to choose the shape you wish to use for the pictogram. This can be changed for each category.
- Drag and drop the shape up to ten times onto any row on the pictogram.
- Press 'reset' to clear and start a new pictogram.

What to do

Find out about the children's favourite pets by a show of hands or by asking round the class. Use the data collected to complete the pictogram. Ask a number of questions relating to the pictogram to check the children's understanding of what the shapes represent. Ask how many shapes would need to be shown if one shape represented two children rather than just one child in the class.

Differentiation

Less confident: ask some children to select their favourite pet, and drag and drop a shape into the correct place on the pictogram.
More confident: ask how many shapes would need to be shown if one shape represented three children rather than just one child in the class.

Key questions

- *What is the favourite pet of the children in the class?*
- *What is the least favourite pet of those selected?*
- *How many children altogether selected a pet?*
- *What is the difference between the most popular and least popular selections?*

labels
Use keyboard to type in title and category labels

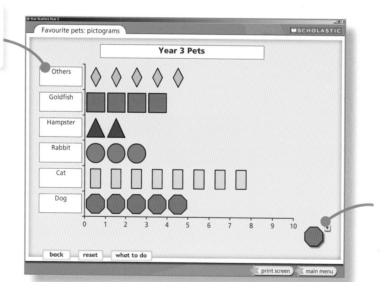

shape selector
Press arrow to choose new shape

Pirates: adding multiples of 10

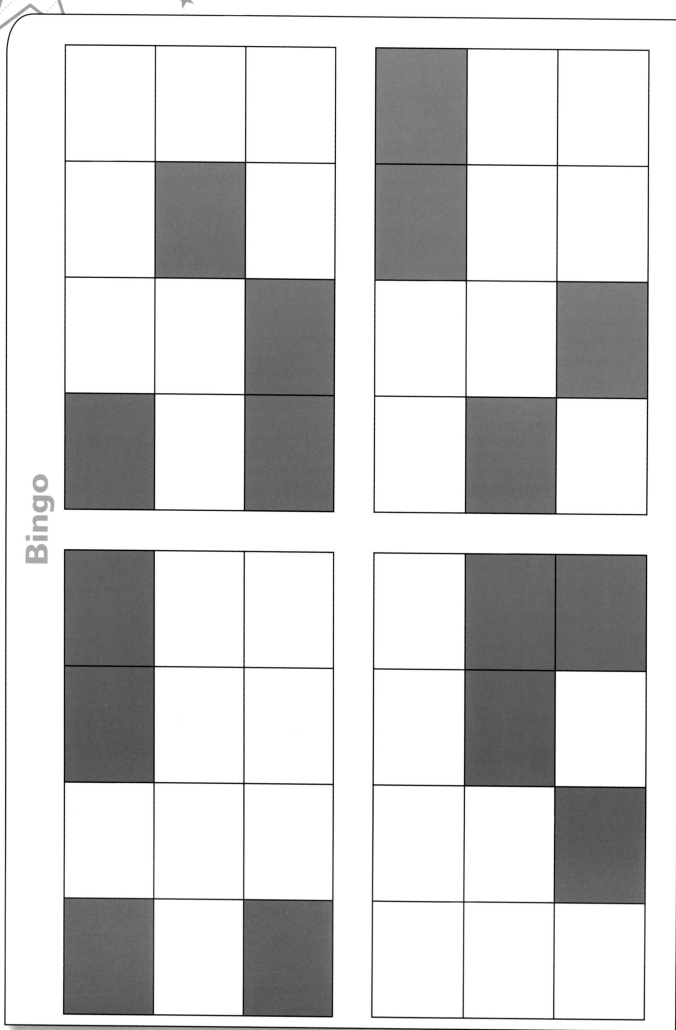

Targets: addition and subtraction

▪ Find the target numbers using the cards below.

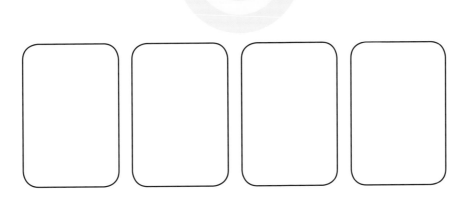

▪ How I worked out the target answer.

_____

_____

_____

_____

_____

_____

_____

_____

Maps and directions

◼ Plan your route using the map below.

Moves

_____

_____

_____

_____

_____

_____

_____

Teacher's name _____

Star Maths Starters diary page

Name of Star Starter	PNS objectives covered	How was activity used	Date activity was used

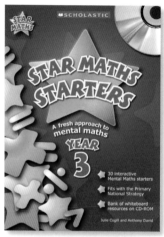

SCHOLASTIC

Also available in this series:

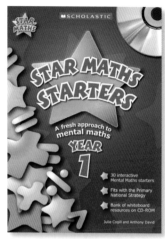

ISBN 978-1407-10007-4

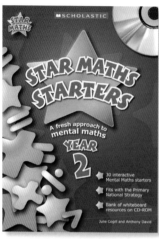

ISBN 978-1407-10008-1

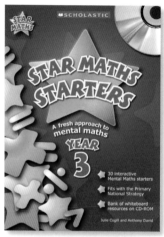

ISBN 978-1407-10009-8

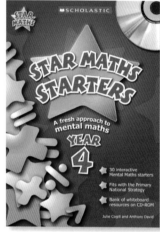

ISBN 978-1407-10010-4

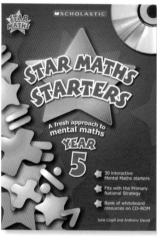

ISBN 978-1407-10011-1

ISBN 978-1407-10012-8

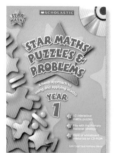

ISBN 978-1407-10031-9

ISBN 978-1407-10032-6

ISBN 978-1407-10033-3

ISBN 978-1407-10034-0

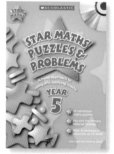

ISBN 978-1407-10035-7

ISBN 978-1407-10036-4

To find out more, call: 0845 603 9091
or visit our website www.scholastic.co.uk